In Brian's first popular book, <u>Mailbox Moo-la, How Real Estate Cash Cows Put Money In Your Mailbox</u>, he focused on the seven years of investing that lead him to support himself and his family solely from the income produced by his rental properties. At the young age of 37, he retired from the day- to-day grind of a 9-5 job. His only job now... to collect the money from his mailbox.

The book received acclaim from his peers. It was recommended as a "must read" in *Commercial Investment Real Estate Magazine*, distributed by the CCIM Foundation, highlighted in Georgia Realtor magazine, and is available on Amazon.com.

In his second book of the series, <u>Mailbox Moo-la...Special Commercial & Land Edition,</u> Brian used his years of representing commercial landlords and tenants, selling multi-family housing, and working land deals to simplify the language of commercial real estate. His conversational style of writing made for an easy read and simplified the commercial and development process for newbies and experts alike.

Now, with the third book of the series, <u>Mailbox Moo-la...Instruction Manual,</u> Brian breaks down these commercial real estate principles into a workbook that can be used to reinforce the reader's use of the terms presented. Real world examples with an answer key help the reader use the skills learned in their individual investment quest.

Take it from a guy who has done it and discover how to:

- Create Consistent Cash Flow
- Cut Your Taxes
- Pick Out the Right Property
- Understand Commercial Real Estate terms
- Get the most out of Leverage and Why it's Important
- Understand the Rezoning and permitting process
- Use 1031 exchanges to build wealth
- Understand the difference between net worth vs. cash flow investing

Brian Patton, CCIM
P.O. Box 2322
Cumming, GA 30028
bpatton@ccim.net
www.MailboxMoo-la.com
770.634.4848

This publication is designed to provide general information regarding the subject matter covered. However, laws and practices often vary from state to state and are subject to change. Because each factual situation is different, specific advice should be tailored to the particular circumstances. For this reason, the reader is advised to consult with his or her own advisor regarding that individual's specific situation.

The author has taken reasonable precautions in the preparation of this book and believes the facts presented in the book are accurate as of the date it was written. However, neither the author nor the publisher assumes any responsibility for any errors or omissions. The author and publisher specifically disclaim any liability resulting from the use or application of the information contained in this book, and the information is not intended to serve as legal advice related to individual situations.

ABOUT THE AUTHOR

Brian Patton, CCIM, has been educated and trained in many aspects of the real estate industry. Previous experience has included land planning large scale master planned golf communities throughout the United States and foreign countries, land planning mixed use commercial developments in the Atlanta Region, and brokering office, retail, restaurant, and land deals in the state of Georgia.

Included with this experience were several years of developing site plans for commercial, industrial and residential developments. A stint in the public sector, as well as zoning and development consulting work, has afforded Mr. Patton a distinct knowledge of the governmental approval process.

Brian also is an author and publishes a semi-monthly column on real estate issues in seven different newspapers with a weekly copy circulation of 120,000+. He speaks as a lecturer to groups on real estate issues and is a past guest with GA Reia, the largest real estate investors association in the country. He just recently finished up a two year stint as the co-host of a popular radio show on investment real estate.

Mr. Patton obtained his Registered Land Planning / Landscape Architecture status in 1993 and his Real Estate licensure in 1993. His designation as a CCIM came in early 2002.

He shares his experiences in investment real estate with readers of this book in hopes that they will someday reach their dreams of mailbox moo-la freedom.

Mailbox Moo-la.com
...How Real Estate Cash Cows put Money in your Mailbox...

INSTRUCTION MANUAL
and
ANSWER KEY

Application of math principles to the concept of owning real estate can provide an investor with a better understanding of how to make money in the real estate market.

Real estate investing can be a rewarding experience. It provides the investor a way to leverage financed funds to increase his or her return on investment; and at the same time, providing a living or working environment for someone's enjoyment.

Investing is giving up something today for something tomorrow. Your job as an investor is to carefully consider all the hurdles to determine how best to use your limited funds to create more funds for the future.

Real estate is the number one generator of wealth in the United States. Most of the first generation millionaires in America received the majority of their wealth through real estate investments. It is with this in mind, that we created this workbook supplement to the mailbox moo'-la book which further enforces real estate principles and math principles simultaneously.

Being a real estate entrepreneur can take many forms, from renting out a house, to owning apartments or retail shopping centers, or developing raw land. With a fundamental understanding of these principles and the math required behind these principles, you will have a solid foundation for understanding the world's greatest wealth generator; and at the same time, honing valuable math

HOW MUCH IS THE RENT?

Determining the value of most real estate starts with calculating the amount of income that can be collected. This can be in the form of rent, late fees, or other income generated by the property. Therefore, it is important to know how to calculate what is called the "gross income." The term "gross income" is defined as, "the total amount of income collected over a period of time prior to subtraction of expenses." The "gross income" will help us in later chapters to determine the value of a real estate investment.

Problem

1. An investor owns a two unit building (called a duplex). Each tenant pays $800 per month in gross rent. What is the yearly gross rent for the entire building?

2. Johns owns a 10 unit apartment complex. Each unit produces $11,000 per year in gross income. What is John's yearly gross income?

3. Emily owns a building with a dog groomer as a tenant. The tenant pays $1,500 per month. What is Emily's yearly gross income?

4. Tammy has purchased a four unit apartment. The units together produce $4,400 per month in gross rents. What is Tammy's yearly gross income?

5. Brian owns twenty five single family houses. The gross yearly rent is $285,000. What is the average gross monthly rent for each home?

6. An investor owns a 20 unit apartment building. One half of the units are one bedroom apartments and the other half are two bedroom apartments. The one bedroom rent is $600 per month. The two bedroom rent is $725 per month. What is the yearly gross rent for the apartment building?

7. John owns three houses and collects rent every two weeks, not every month. The rent is $300 every two weeks. What is his gross yearly rent on the three houses?

8. Penn owns a shopping center with three stores: a Nail Salon, a Sandwich Shop, and a Barber Shop. Each business pays $2,000 per month in gross rent. What is the gross yearly rent for the shopping center?

9. Penn also owns a small apartment complex. There are 30 units and the gross monthly income equals $30,000. He also owns a Laundromat on the property that generates another $25,000 per year in gross income. What is the gross yearly income from the property?

10. Brian wants to purchase a small office building that is rented to Chandler Chiropractic & Associates. Dr. Chandler pays $3,000 per month. However, Brian has found another tenant, Dickenson Attorneys at Law, that will pay $1,500 every two weeks. Which tenant would generate the most gross yearly income for Brian?

IS IT A GOOD DEAL?

Most rents for commercial buildings, or apartments, are determined based on a cost per square foot basis. In other words, the rent is determined by taking an amount per square foot and multiplying it by the square foot of space someone leases. Since tenant spaces are all different sizes, this method is an easy way for potential tenants to compare the different spaces. It's a way for the tenant to figure out if they are getting a good deal. While different areas of the country and different landlords use different calculations, we will use a standard per year rent calculation. For instance, a property that rents for $1 per month per square foot, would also rent for $12 per year per square foot. Unless otherwise noted, our rents will be stated in the form of the per year rent.

As an example, a retail strip center charges $20 per square foot per year in rent. A 1,000 square foot space then would rent for $20,000 per year. $20 x 1,000 square feet = $20,000

PROBLEMS

1. A hair salon pays $15 per square foot for 2,000 square feet of space. What is their yearly gross rent?

2. A barber shop pays $40,000 a year in rent. If their rent is $20 per square foot, how many square feet of space are they renting?

3. Jorge owns a building in downtown and rents the building to a delicatessen. The store is 4,000 square feet in size and the tenant pays $25 per square foot. What is Jorge's yearly gross rent?

4. A local pharmacy pays $50,000 a year in rent. If there space is 2,000 square feet, what is the per square feet costs of the space?

5. A bakery is looking for new space. If the most they want to spend is $30,000 per year and they need 1,500 square feet, what is the most they can pay per square foot for the space?

6. A new grocery store is looking for a new location. They have found two spaces that they like. Space A is 50,000 square feet and rents for $15 per square foot. The other, Space B, is a 45,000 square foot space and rents for $18 per square foot. Which space will be cheaper rent for the grocery store?

7. Brian owns a retail shopping center with four tenants. Two tenants rent 2,000 square foot each, and two tenants have 5,000 square foot each. If Brian requires $20 per square foot for each tenant, what is the gross monthly rent for the entire building?

8. A bank just foreclosed on an office building that has a tenant that is paying $36,000 per year in gross rent. The building is 5,000 square feet. How much per square foot is the tenant paying in gross rents?

9. The bank, from problem #5, has offered to let the tenant out of the lease. The tenant has found space in a similar building, Building B. The rent is $10 per square foot and the building is 4,000 square feet. Which gross yearly rent is cheaper for the tenant?

10. Tammy owns an apartment complex. Her average rents are 95 cents per square foot per month. If the average apartment is 1,000 square feet. What is the gross yearly rents for 20 of these units?

WHERE DID MY TENANTS GO?

As a landlord, once you collect the gross rents, you must deduct your expenses from this rent. These expenses are known as operating expenses. Expenses can be property taxes, property insurance, maintenance and repairs, or vacancy. The bane to any landlord's existence is vacancies. Nothing will cost you more money, more headaches, and more sleepless nights than having vacancies. But, you'd better "cowboy up," as they say out West. Because, if you own a rental property, then you will have vacancies. We refer to this inevitable situation as the "vacancy rate." The "vacancy rate" is a percentage of time that you will have vacancies over a year period. It can range anywhere from five percent to twenty percent. Vacancy rate is similar to the unemployment rate...it's the amount of time that your units will be unemployed. Typically, as with the unemployment rate, five percent is about the best you can expect from the vacancy rate.

Now, I used to think when I owned just one unit, that this vacancy rate made little sense. My logic was that either it's vacant or not...so the vacancy was either zero percent or one hundred percent. But, as I began to gain more experience as an investor, I learned that over a lengthy period of time; say five years, that the vacancy rate for my units tended to settle right in at five percent per year. A large part of this is due to tenants moving out and how long it takes to get a unit ready to rent again. Even if you had a great tenant that left your unit in perfect shape, there's still a turnaround time. A simple five percent vacancy means that out of the 52 weeks of the year, that it would be vacant roughly two and a half weeks out of the year. This doesn't leave a lot of time to clean the unit, advertise, show it to a few people, sign a lease, and work around the tenant's move-in date.

When you are calculating income, you should always use this vacancy rate in your calculations.

As an example, if your gross income is $10,000 per year, and there is a 5% vacancy rate, then you should expect to be out of pocket at least $500 in vacancy over the year period. $10,000 x .05 = $500

PROBLEMS

1. Joshua has two houses that he rents for $1,000 per month each. He knows that he must apply a 5% vacancy rate to his gross rent. What is the dollar amount of this vacancy per year?

2. Barbara is a new investor and has one house that she rents to a young couple for $950 per month. She doesn't apply a vacancy rate to the income, because it's never been vacant since she bought it six months ago. Should she apply a vacancy rate anyway? Why or why not?

3. Francisco's apartments generate $35,000 per year in gross rents. Francisco has learned that a 5% vacancy rate should be applied to his gross rents. What is the dollar amount of this vacancy rate calculation?

4. Emily has purchased an office building to rent to an accountant. Her tenant, the accountant, pays $40,000 per year in gross rents. If the vacancy rate is 10% in this area, what is her net income after deducting the amount of the vacancy?

5. John's gross income is $20,000 on an office building he owns. The vacancy rate for this area is generally around 5%. What is the vacancy costs in dollars for John? If John's building becomes vacant, does that change the vacancy rate for John? Why or why not?

6. No matter how hard she tries, Bekah cannot keep her apartments rented all the time. She has 10 apartments that rent for $800 each per month. She knows that she should apply a vacancy rate because of this problem. If she applies a 5% vacancy rate, what is the dollar amount of this vacancy?

7. Pendleton owns a building with two restaurants in it. A sandwich shop pays $30 per year per square foot in rent and has approximately 1,000 square feet. The pizza parlor pays $25 per year per square foot in rent and has approximately 1,500 square feet. Pendleton knows that, since restaurants are difficult businesses to operate, they have pretty high vacancy rates. He decides to use a 12% vacancy rate. What is the dollar amount of this vacancy calculation?

8. In the example above, what is Pendleton's income after subtracting his vacancy expense from the gross rents?

9. Emily has an interior design studio in a renovated downtown building. She rents space from a local business man for $18,000 per year. If he applies a 7% vacancy rate, what is the income after subtracting this vacancy rate?

10. Josh owns a computer programming business and rents space for $30,000 per year. His landlord applies a 10% vacancy rate to this building. What is the dollar cost of this vacancy rate?

IS THERE ANY MONEY LEFT?

Other than vacancy, there are other costs that are also called "operating expenses". The most common of these are property taxes, property insurance, and maintenance costs. To eventually determine the value of a real estate property, which we will do in a later chapter, we must calculate the income after subtracting these expenses. These operating expenses include the items below:

1. Real estate and personal property taxes
2. Property insurance
3. Management expenses
4. Repairs and maintenance
5. Utilities
6. Accounting and legal services
7. Vacancy

To eventually determine the value of a property, we will calculate the net operating income (NOI). The (NOI) net operating income is a very common term used by real estate investors. It refers to the yearly income you have on the property after subtracting your expenses. One important expense that we do not include in calculating the NOI is the mortgage payment. The reason is the value will not be affected by the amount of the mortgage payment, since one person may pay cash and one person may put down only ten percent. The value is strictly based on the income minus the expenses.

To determine this net operating income, we simply take the gross income and subtract the operating expenses, including the vacancy expense. An equation that can be used for this calculation is:

Gross rents (GR) minus the operating expenses (OE) equals the net operating income (NOI).

$$GR - OE = NOI$$

PROBLEMS

1. An investor collects $18,000 per year in gross rents. His operating expenses equal $3,000 per year. What is his net operating income?

2. John's investments create $5,000 per month in gross rents. If his yearly operating expenses equal $5,500, then what is his monthly net operating income?

3. Bekah owns a dog grooming business. She rents a small building for $12,000 per year. The owner of the building has operating expenses of $1,675 per year. What is the owner's net operating income?

4. Bob owns a building and rents it to a car wash. The car wash pays $50,000 a year in gross rent. Bob's operating expenses are $5,000 per year and his mortgage debt costs him $10,000 per year. What is Bob's net operating income?

5. Royce owns a house that he converted into an office. A potential tenant has offered him $995 per month in gross rent. The tenant agrees to pay the property taxes of $1,500 per year. The other expenses are Royce's responsibility. Royce's expenses total an additional $1,700 per year including the vacancy rate. What is Royce's net operating income?

6. Brian owns 18 apartment units that generate $216,000 gross rents per year. His property taxes are $45,000 per year, his insurance expenses are $9,000 per year, his maintenance costs is another $5,000 per year, and the vacancy rate of 5% costs him an additional $10,800 per year in costs. What is Brian's net operating income?

7. Doug owns a retail center with four spaces of 2,000 square feet each. They generate $20 per year per square foot in gross rents. His operating expenses, including vacancy, are $5 per square foot per year. What is his net operating income?

8. Barry recently purchased a building with an assisted living facility as a tenant. The property taxes are $2,000 per year, the property insurance is $1,700 per year, and the maintenance costs another $4,000 per year. The tenant pays $60,000 a year in rent. Using a vacancy rate of 10%, what is Barry's net operating income (NOI)?

9. Amanda bought a house to rent in a small town in Mississippi. The monthly rent is $600 per month. Her expenses equal $1,800 per year, not including vacancy rate. Vacancy rate for this area averages 8%. What is Amanda's net operating income?

10. Paul recently rented a small 1,500 square foot location for a barbeque restaurant. His landlord agreed to accept $10 per square foot per year as the rent. The landlord's total expenses, including vacancy, are $250 per month. What is the landlord's net operating income?

WHAT IS IT WORTH?

Next we will take the NOI (net operating income) and determine the value of a real estate investment using the "capitalization rate." The cap rate, for short, is a rule of thumb that we use to compare investments. A simplified definition for cap rate is: it is the return on your investment if you paid cash for the property. Cap rate is expressed as a percentage. Virtually all real estate values are determined by this method.

The formula to determine the value using the cap rate is:

value = NOI / cap rate.

To determine the cap rate, the formula is:

Cap rate = Value / NOI

And to determine the NOI, the formula is:

NOI = Value x cap rate

Let's look at an example. If you paid $100,000 in cash for a real estate property and rented it for $10,000 in income per year; then you would have a 10 percent return on your investment. This return on your investment is also called the cap rate.

Value = $100,000
NOI = $10,000
Cap rate = value / NOI
$$= \$100,000 / \$10,000 = 10$$
Cap rate = 10%

The actual cap rate that you use is entirely up to you as an investor. It is just a rule of thumb that you will learn from experience. Cap rates can range from 4% to 20%, depending upon the investor's beliefs about the market and the property they want to buy. In some ways, this method of determining the value is very subjective. However, as you begin to invest in real estate, you will start to form your own opinions about what cap rate you would like to receive from your investment.

Note: As with our other calculations so far, we do not consider the amount of mortgage payments. This mortgage debt is not important to the value of the property.

PROBLEMS

1. Dr. Lee has purchased a retail building with a NOI (net operating income) of $50,000. He knows from experience that a fair cap rate would be 10% for retail buildings in this area of the country. Based on that assumption, what is Dr. Lee's value of the real estate?

2. Judy owns an apartment complex with 20 units. The NOI (net operating income) is $150,000. Using a cap rate of 12%, what is the estimate of value?

3. Vicky owns 10 single family houses with $120,000 per year in NOI (net operating income). If she sells these houses to another investor based solely on the cap rate formula. What would her value be based on a 13% cap rate?

4. Brian owns a shopping center with a grocery store and three other tenants. He believes his building is worth $5,200,000 based on a cap rate of 8%. What is his NOI (net operating income)?

5. John's real estate investments create a NOI (net operating income) of $150,500 per year. His mortgage debt is $30,000 per year. If he believes his investments are worth $1,505,000, what cap rate is he using to determine this value?

6. One of Tom's properties has a NOI (net operating income) of $50,575 per year. He knows that cap rates for this type of property range from 8% to 9%. What would be the value of the property based on both of these cap rates?

7. Based on the answer above, if Tom wants to sell the property, which cap rate would he prefer?

8. If John wants to buy the property from Tom in the question above, which cap rate would he prefer? Why?

9. Obviously, there is a correlation between the value and the cap rate. If the value goes up, what happens to the cap rate? If the value goes down, what happens to the cap rate?

10. Joe, the Mayor of a small town in Georgia, owns a downtown building. He has calculated that the value of the building is $590,000, based on a 10% cap rate. What would the value be if he used a 9% cap rate instead?

HOW MUCH WILL I MAKE?

Now we have determined something very important: the value of a property. Next, we will look at how much money we will actually make, called the "cash flow". When you go to a bank to borrow money to purchase something this is called financing. Using a credit card to purchase clothing, food, or a computer is using financing. While it's possible to purchase real estate without financing, most real estate investors will use some form of financing.

Financing is the use of borrowed funds for a purchase. Financing is also called leverage, because you are using borrowed funds to "leverage" a purchase. Using leverage increases your power by using other people's money, namely your banker's money.

One advantage of using this leverage is obviously you can purchase more expensive property. Without financing, most of us wouldn't be real estate investors. Let's face it, most people can't whip out $200,000 to buy a rental house, but some of us can find $10,000 or $20,000.

It also allows you to keep your cash to purchase even more properties. This helps reduce your risk since it allows you to diversify your real estate investments. If you have one or two units in numerous subdivisions or in different cities, your risk is spread out. If a subdivision goes down in value, the rest of your properties pick up the slack. If a particular area decreases in value because of an unsightly landfill or deterioration of a school district, then you're protected by the rest of your real estate investments.

But, my favorite reason for using leverage is that it increases your return on investment. Let's look at an easy example.

Let's say your long lost uncle passes away and gives you $100,000. What do you do with it? Maybe buy some real estate?

Let's say a $100,000 house rents for $1,000 per month. If you pay cash for the house, then your cash flow is roughly $12,000 per year, or a 12% return on investment. Not bad.

But, if you use financing, then you put $10,000 down and have a $90,000 mortgage. With a 7% interest rate, your payment would be around $600 per month. You would make $400 per month ($1,000 in rent minus $600 mortgage payment) You would make $4,800 per year. Remember, your initial down payment was only $10,000, yielding a whopping 48% return! That's four times the rate of return from paying cash, as in the example above. Now, that's a huge return and that's why I get excited about investing in real estate and using my banker's money .

To take this example further: Since you still have $90,000, let's say you buy 9 more houses. So, now you have a total of ten houses, each producing and income of $4,800 per year. That's a total of $48,000 per year on a $100,000 investment. Still a 48% return and a lot of money in your pocket. That's why I love leverage!

There are generally two types of loans: fixed loans and variable loans. Fixed rate loans will not vary over time. Their interest rates are "fixed"; hence the name. The payment on a fixed rate loan will be the same in the future as it is now. Variable rates will vary, or change, over the years. Their rates generally start out low; as a teaser, then go up as the loan gets older.

Typically, if you buy a house or duplex, you get a 30 year loan, which means if you pay the same monthly payment every month for 30 years then the loan is paid off. Some properties, such as commercial buildings and apartments, may use a shorter time period, such as 20 years.

When someone talks about paying the mortgage payment, they are usually talking about the common four aspects of a loan:

Principal
Interest
Property taxes
Property insurance

Principal is the amount of money you have borrowed. Interest is the amount of money the bank makes because you've borrowed money from them. Most loans contain both principal and interest. However, sometimes a bank will allow you to do an "interest-only" loan. That means, that you don't pay any of the principal amount; you just pay an interest payment every month.

When you finance a real estate investment, sometimes the bank will require you to "escrow" the property taxes and property insurance. These are two very important items that the bank wants to make sure are paid. So, they will take the yearly amount of these two items divide them by 12 and require you to pay this amount every month. However, if you remember, we accounted for these two items in a previous chapter. They are accounted for in the expenses section where we determined NOI (net operating income). Therefore, when we talk about mortgage payments in this chapter, we do not include taxes and insurance, as they have already been subtracted.

CASH FLOW

If we use financing in buying real estate, then we can take our NOI (net operating income) and subtract the yearly mortgage payment. Remember, NOI (net operating income) is a yearly number. But, most people think of the mortgage payment as a monthly expense. So, you may have to multiply your monthly mortgage payment by 12 months to calculate the yearly mortgage payment.

So, Cash Cow lovers, here is the number we've been waiting for. This is our "cash flow." This is the amount of money we get to put in our pockets.

Here's the formula:

NOI – yearly mortgage payment = yearly cash flow

PROBLEMS

1. Tommy rents a building to a television repair shop. Tommy's NOI (net operating income) is $11,000 per year. His yearly mortgage payments total $8,000. What is his yearly cash flow?

2. Audrey Ann wants to purchase a building. Her tenant, a beauty salon, will be paying rent. Audrey Ann's NOI (net operating income) will be $15,500 per year. Her mortgage payment will be $11,655 per year. What is Audrey Ann's yearly cash flow?

3. Brian owns a duplex apartment building. His NOI from the building is $24,000 per year. His monthly mortgage payment is $800 per month. What is Brian's yearly cash flow?

4. Paul is losing money on his investment. His NOI (net operating income) totals $12,000 per year on a small, two bedroom house he owns in the city. If his mortgage payment totals $900 per month, what is his cash flow? Is he losing money?

5. Paul must be pessimistic, because he also thinks he's losing money on a three bedroom house he owns in the suburbs. His NOI (net operating income) is $15,000 per year. His mortgage payment is $1,400 per month. What is Paul's cash flow? Is he losing money?

6. John's mortgage payment is $1,200 per month. $200 of that goes to property taxes and insurance. His real estate investment produces $15,000 a year in NOI (net operating income). What is John's cash flow?

7. Jeremiah owns a building with a car wash as a tenant. His monthly mortgage payment consists of $500 in principal payments, $200 in interest payments, $300 in property taxes, and $75 in property insurance. The car wash produces $22,555 a year in NOI (net operating income). What is Jeremiah's cash flow?

8. Sally wants to purchase a house to rent out. Her initial calculations indicate a NOI (net operating income) on House A to be $10,555 per year. House B will generate a NOI of $11,675 per year. Assuming her mortgage payment on either property will be $750 per month, what is the cash flow on House A and what is the cash flow for House B? Which is the better investment?

9. Francisco has determined that an investment in a 25 unit apartment complex will provide a NOI of $275,000 per year. If his principal payment is $60,000 per year, his interest payment is $173,000 per year, and his taxes are $5,000 per month, what is his yearly cash flow?

10. Ralph's banker insists that his mortgage payment include property taxes and property insurance. Ralph's principal and interest is $895 per month. His property taxes and property insurance totals $300 per month. If Ralph's NOI is $30,600, then what is his yearly cash flow?

WHAT DO I DO NOW THAT I'VE MADE ALL THIS MONEY?

Now, that you've made money on your real estate investment, you will have to pay some taxes. However, real estate is one of the best, if not THE best, tax shelters available. The term "tax shelter" has some negative connotations, because of frequent misuse. But, the great part of real estate tax law is that it allows you to not pay taxes on some of your income. In other words, you get to "shelter" some of this income from taxes.

Real estate is a legal tax shelter that you can use to reduce your personal income taxes. The government agency that is in charge of deciding how much in taxes we pay is called the IRS (Internal Revenue Service). The primary tax shelter for real estate used to be called depreciation by the IRS, but is now referred to as "cost recovery". I'll use depreciation since it's an easier term to use. Depreciation means that because buildings get older and require more and more maintenance over the years, then their value declines because it takes more money to keep them repaired properly. The value of the land is not depreciated since land cannot depreciate. It remains virtually the same over time.

The beauty of the tax law is that while your real estate investment is actually appreciating, the federal government allows you to pretend that it's depreciating in value. Residential property is depreciated over 27 ½ years. In other words, the IRS allows you to pretend that the buildings on the property will be worth nothing in 27 ½ years. So, every year you'll get to depreciate the value the same amount until 27 ½ years later. Commercial is depreciated the same way except over 39 years.

For an easy example, let's look at a rental house that's worth $100,000. Since land doesn't depreciate, we only consider the value of the house on the property. So, in this example, the actual house may be worth only $80,000 and the land that is sits on may be worth $20,000. So, you simply divide $80,000 by 27 1/2. This comes to an amount of $2,909 per year of depreciation value.

This is to say that the IRS assumes that the building depreciates the same amount every year for the next 27 ½ years, an amount of $2,909. This amount comes off of your personal taxes, but only for investment properties. You are not allowed to deduct deprecation in your principal residence...the one you live in.

So how does it save you money? The amount that this saves you in tax money depends upon your income tax bracket. If you are in the 35% tax bracket, then in this example, your personal taxes would be reduced by $1,018 per year. ($2,909 x 35%).

PROBLEM

1. A house that is valued at $80,000 is allowed depreciation of $2,909 per year. How many years will the investor be able to use this depreciation amount?

2. In the example above, if your personal income tax percentage is 21%, then how much will you save in taxes per year?

3. Wynona has a commercial building that is valued at $350,000 and the land is worth $75,000. Using a depreciation amount of 39 years, what is the yearly depreciation amount?

4. If a house and a commercial building are valued at the same amount, then which one will produce the most yearly amount in depreciation?

5. Benjamin has four houses that he rents out that he has owned for 11 years. Each building is valued at $120,000. The land is worth $35,000 per each lot. How many years should Benjamin use to determine the depreciation?

6. If Benjamin in question #5 depreciates the buildings, what is the yearly amount of depreciation value?

7. If Benjamin in question #5 is in the 46% tax bracket, then how much will this depreciation save him in personal taxes?

8. In question #7, how many years will Benjamin be able to save this amount of money on his personal taxes?

9. If Donald owns a commercial building that is valued at $480,000, not including the land value, then what is the amount of deprecation per year?

10. From Question #9, what is Donald's personal tax savings, if he is in the 36% tax bracket?

CAN I SAVE EVEN MORE ON TAXES?

There are several more deductions that you can take when using real estate as an investment. One of those items is the payment of interest. As we discussed in a previous chapter, mortgages are usually made up of two main parts: principal and interest. The interest part of this payment is deductible on your personal taxes. This can result in additional tax savings.

For instance, if you have a loan of $100,000 at 7% interest, then the monthly payment is $665.30. The yearly principal amount will be $1,015.81 and the yearly interest payments total $6,967.82. (we calculate this on a financial calculator)

This yearly interest payment will be deductible from your personal taxes. If you are in the 25% federal tax bracket, then your personal taxes will be reduced by $1,741.95 per year ($6,967.82 x .25). Add this to your depreciation deduction that we discussed in the previous chapter, and you are well on your way to reducing your taxes quite a bit.

PROBLEM

1. If John pays $5,125 per year in interest on his investment and is in the 25% federal tax bracket, what is his personal tax savings?

2. Tammy owns two houses. She pays $6,324 per year in interest on one house and $5978 per year in interest on another house. Both houses are investment properties. What is her total in interest that she pays yearly?

3. John owns a commercial building with three tenants. The interest on the building is $25,435 per year. If John is in the 35% federal tax bracket, what will be his tax savings generated by this interest?

4. If Tim saves $3,500 per year in taxes just based on his interest deduction, how much interest does he pay if he is in the 25% tax bracket?

5. How much in taxes would Tim save if he were in the 36% tax bracket?

6. If John pays $8,967 per year in interest, how much in taxes will he save if he is in the 25% tax bracket?

7. Robert pays $2,539 in principal and $17,419 in interest per year on a four unit apartment that he owns. Can he deduct the principal amount that he pays? Why or why not?

8. From #7 above, what is Robert's tax savings based on a 36% tax bracket?

9. What would Robert's tax savings be if he were in the 25% tax bracket?

10. If Samantha saves $3,905 per year in interest tax deductions and she is in the 25% tax bracket, then what is the amount of interest she pays?

HOW MUCH OF THIS CASH DO I GET TO KEEP AFTER TAXES?

In a previous chapter, we calculated "cash flow" created by our cash cows. In subsequent chapters, we learned of some of the tax benefits of real estate. Now, we will calculate our "taxable income." In other words, how much of this cash is subject to taxes. And, how much do I get to keep after paying taxes and after finding out my tax savings.

Let's remember how we got to this point from previous chapters:

      Gross income
-   Operating income
      NOI (net operating income)
-   Mortgage payments
      Cash flow
-   Tax deductions (deductions and interest)
      Cash flow after taxes

Let's look at the following example. Rick purchased a building that leases for $30,000 per year. His expenses, (i.e. $3,000 property taxes, $1,500 vacancy expense (5%), $2,500 maintenance expenses, and $700 legal expenses) total $7,700.

Using our formula: Gross income − operating expenses = NOI
Rick's NOI (net operating income) equals: $30,000 - $7,700 = $22,300

Rick's mortgage payment: $19,160 per year.

Using our formula: NOI – mortgage payment = cash flow
Rick's cash flow is: $22,300 – 19,160 = $3,140

As we learned in a previous chapter, we can determine Rick's property value based on the NOI (net operating income). Remember, we use a cap rate to determine this. So, if we use a 9% cap rate, we will find his value using the following formula:
Value = NOI / cap rate
Value = $22,300 / .09 = $247,777

We can estimate the value of the building on the property by multiplying the value of Rick's property by 80%. Remember, the land cannot be depreciated, only the building. This 80% is just a rule of thumb we will use to help us with this calculation.

So, Rick's building is valued at: $247,777 x .80 = $198,222

Rick's building is a commercial building, so we are allowed to depreciate the value the same amount every year for the next 39 years. (27 ½ years if it were residential)

$198,222 / 39 years = $5,082.62 per year

Also, we can deduct the interest that Rick pays on his mortgage debt. We find from our financial calculator that the interest is $16,722.77 per year.

So, Rick is allowed to deduct $5,082.62 + $16,722.77 = $21,805.38

TAXABLE INCOME

Now, we can take what we learned in the previous two chapters on depreciation and interest deductions. The taxable income is determined by subtracting these two from the NOI as in the following:
NOI – depreciation – interest payments

Rick's taxable income =
$22,300 (NOI) - $5,082.62 (depreciation) - $16,722.77 (interest) = $494.61

Here's the GOOD PART:

Remember, Rick's cash flow was $3,140. But, he only has to pay taxes on $494.61. If he is in the 25% tax bracket, then he only has to pay $123 ($494.61 x .25) in taxes.

He made $3,140 and only paid $123 in taxes. This equals only 3.9% of his actual cash flow in taxes.

So, Rick's after tax cash flow is: $3,140 – $123 = $3,017

PROBLEMS

Answer the following problems based on Rick's scenario above:

1. What was Rick's Gross Income?

2. What was Rick's operating expenses?

3. What is the formula to determine Rick's NOI?

4. If Rick were in the 35% tax bracket, what would his tax bill be?

5. If Rick were in this 35% tax bracket, what would his after tax cash flow be?

"A Bird in the Hand"

You've probably heard the old saying, "A bird in the hand is worth two in the bush." This saying also applies to real estate investing. To refer to this concept, we use the term "future value." You will either receive income today, tomorrow, or in the future. It's important to understand this concept because most of your real estate income will come in the future. Therefore, you must know what that income is worth today. In my real estate courses, I've taught that a dollar today is worth more than a dollar tomorrow. This is due to two reasons:

1. Number one is the affect of inflation on the value of a dollar. Inflation is defined by Wikopedia.com as "a rise in the general level of prices of goods and services in an economy over a period of time." Inflation decreases the value of your dollar every day. If it takes one dollar today to buy a loaf of bread and in five years it costs two dollars, then the value of your dollar has been cut in half. Therefore, **your dollar is worth more right now than it will ever be.**

2. The second reason is the idea of risk. There is some level of risk in any investment that you won't get paid in the future. If you have the dollar in your pocket, there is no risk. You already have it and don't have to worry about receiving it later.

So, because most of your cash flow will happen in the future, it's important to know what that money is worth right now. Using a financial calculator, you can determine the value of future dollars based on what's called a "discount rate." It's also referred to as a "costs of capital rate." This is merely the interest rate that you think you would make on the money in some other investment if you didn't put it in this one. For instance, let's say to buy a property you have to put $10,000 as a down payment. Obviously, you have lost some opportunity with that money. You could have put it in the bank, you could have bought some stock, or some other investment. Your opportunity costs, then, must be determined and put into this equation. This interest rate is entirely based upon your individual circumstances –it's entirely your opinion or your experience that determines what "discount rate" you will use.

We can do some simple calculations to help us understand "future value." While most of these calculations are normally completed on an advanced financial calculator, I think some simple calculations will help us understand the concept.

Let's suppose that you've invested in a piece of land. It's a small two acre tract of land in a growing area that you bought a few years ago. A developer wants to build a day care on the property. Since the developer has to get the property rezoned to allow a day care site, he knows that there is some risk that he may not get it rezoned. He also knows that it will take possibly a year to have it rezoned.

He offers you the following: He'll either buy your property now for $75,000 before the rezoning OR he will rezone it and then pay you $85,000 a year from now. Which one should you do?
The first thing you must determine is your "costs of capital." As I mentioned above, what kind of interest rate can you expect to get from your investments is your "costs of capital." For this example, let's say that you're pretty good at investing in the stock market and have decided to use this money from the sale of the property to the developer to invest in stocks. Your usual return in the stock market has been 10%.

So, if you accepted the $75,000 from the developer, then the value of this money in one year would be:
$75,000 x .10 = $7,500
$75,000 + $7,500 = $82,500

If you wait a year, then you would receive $85,000 from the sale. If you sell it now, then you would receive $75,000, but this amount would be worth $82,500 a year from now. So this calculation helps you better understand the developer's offer. You know the offers are both much closer in value than they originally appeared. Now, you have a tool to make a better decision.

PROBLEMS

1. In the example above, what would be the future value of $75,000 if you used a 15% costs of capital interest rate?

2. In the example above, what would be the future value of the $75,000 if you used a 5% costs of capital interest rate?

3. In the example above, in one year, if the future value of the $75,000 equals $85,000, which deal would you take?   Why?

4. John has an offer from a developer to buy his land. The developer has offered him $250,000 OR $272,000 a year from now. Using a 10% costs of capital, what would the $250,000 be worth in one year?

5. In #4 above, if John uses a 15% costs of capital, what would the $250,000 be worth in one year.

The following two real life scenarios will use all the skills we have learned so far.

SCENARIO #1

Ralph has leased a building for his new bank. The building is 4,000 square feet in size and the yearly rent rate is $16 per square foot. The owner of the building, Brian, expects $5,000 a year in operating expenses. Brian's mortgage debt on the building is $44,472 (This includes principal of $8,023 and interest of $36,449). Brian uses a 9% capitalization rate to determine value of his real estate investments and he is in the 35% federal tax bracket. The building is commercial and depreciation is based on 39 years.

1. What is Ralph's gross yearly rent?

2. What is Brian's NOI?

3. What is the value of the building?

4. Assuming the improvements on the land are 80% of the value of the entire property, what is the depreciation deduction amount?

5. What is Brian's cash flow before taxes?

6. What is Brian's taxable income?

7. What is Brian's cash flow after taxes?

8. What is Brian's cash flow if no taxes are owed?

9. Would Brian's yearly depreciation deduction be higher or lower if this building were considered residential?

10. If Ralph renegotiated the rent to $14 per square foot, what would Brian's cash flow before taxes be?

SCENARIO #2

Mr. Park Ko has purchased a multi-tenant building with three tenants. A Starbucks occupies 1,500 square feet, a hair salon is in 2,000 square feet, and a barbeque restaurant is occupying 2,600 square feet. The restaurant pays $25 per square foot, and the other two tenants pay $20 per square foot. Yearly operating expenses are $20,545. Cap rates for this type of investment are usually around 10%.

Mr. Ko has asked a commercial real estate broker to help him sell the property. However, he doesn't know how much to ask for it. He knows a similar building to his sold for $250 per square foot and believes that is a good comparable.

1. What is the yearly gross rent generated from all three tenants?

2. What is the NOI?

3. Given the parameters above, what is your estimate of value?

4. Do you consider the comparable sales price? Why or why not?

5. Using 80% to value the structures and a 39 year depreciation, what is Mr. Ko's yearly depreciation deduction?

Answer key

HOW MUCH IS THE RENT?

1. $19,200

   $800 per month x 2 units x 12 months = $19,200

2. $110,000

   10 units x $11,000 = $110,000

3. $18,000

   $1500 x 12 months = $18,000

4. $52,800

   $4,400 x 12 = $52,800

5. $11,400

   $285,000 ÷ 25 units = $11,400

6. $159,000

   There are 20 units. Divide the units in half (20 ÷ 2 = 10) and apply the $600 rent to 10 of the units, and apply the $725 rent to the other 10 units.

$600 x 10 units = $6,000 per month
$725 x 10 units=$7,000 per month
$6,000 + $7,000 = $13,000 per month x 12 months = $159,000

7. $7,800

   $300 per month x 26 payments = $7,800
   (Paying every two weeks actually increases rents by two payments, instead of 12 monthly payments, there are 26 payments)

8. $72,000

   $2,000 monthly rent per tenant x 3 tenants = $6,000 per month total income
   $6,000 per month x 12 months = $72,000

9. $385,000

   $30,000 per month x 12 months = $360,000 per year income on the apartments
   $360,000 per year income + $25,000 per year income on the laundromat = $385,000

10. Dickenson Attorneys at Law will generate the most yearly income

    Dr. Chandler's rent $3000 per month x 12 = $36,000 per year
    Dickenson Attorneys rent $1,500 every two weeks x 26 payments (the number of two week payments in a year) = $39,000 per year

IS IT A GOOD DEAL?

1. $30,000

   $15 x 2,000 square feet = $30,000

2. 2,000

   $40,000 per year ÷ $20 = $2,000

3. $100,000

   4000 sf x $25 = $100,000

4. $25

   $50,000 ÷ 2,000 = $25

5. $20

   $30,000 ÷ 1,500 = $20

6. Space A will be the cheapest rent

   Space A   50,000 x $15 = $750,000 yearly rent
   Space B   45,000 x $18 = $810,000 yearly rent

7. $280,000

    Two tenants x 2,000 square feet = 4,000
    Two tenants x 5,000 square feet = 10,000
    4,000 + 10,000 = 14,000 square feet
    14,000 square feet x $20 per square foot = $280,000

8. $7.20

    $36,000 ÷ 5,000 square feet = $7.20

9. The rent is cheaper in the existing building.

    Existing rent is $36,000. Building B rent is $40,000 per year ($10 per square foot x 4,000 square feet)

10. $228,000

    95 cents per month (.95) x 12 = $11.40 per year
    $11.40 x 1,000 square feet per unit = $11,400 per unit
    $11,400 per unit x 20 units =$228,000
    Optional solution:
    .95 x 1,000 square feet per unit = $950 per unit per month
    $950 x 12 months x 20 units = $228,000

WHERE DID MY TENANTS GO?

1. $1,200

    2 houses x $1,000 = $2,000 per month
    $2,000 x 12 = $24,000 per year
    $24,000 x .05 = $1,200

2. She should apply a vacancy rate. Vacancy rates are average vacancies over an extended period of time. Six months is not long enough to determine the application of vacancy rates.

3. $1,750
    $35,000 x .05 = $1,750

4. $36,000
    $40,000 per year x 10% (.10) = $4,000 vacancy expense
    $40,000 - $4,000 = $36,000

5. $1,000 Johns' vacancy rate doesn't change if his building becomes vacant. Vacancy rate is a long term measurement of vacancy.

    $20,000 x .05 = $1,000

6. $4,800

    10 apartments x $800 per month = $8,000 per month
    $8,000 x 12 months = $96,000
    $96,000 x 5% (.05) = $4,800

7. $8,100

Sandwich shop: $30 x 1,000 square feet = $30,000 gross rent
Pizza parlor: $25 x 1,500 square feet = $37,500 gross rent
$30,000 + $37,500 = $67,500 total gross rents
$67,500 x .12 = $8,100

8. $59,400

Total gross rents = $67,500 - $8,100 = $59,400

9. $16,740

$18,000 x .07 = $1,260 vacancy expense
$18,000 - $1,260 =$16,740

10. $3,000

$30,000 x .10 = $3,000

IS THERE ANY MONEY LEFT?

1. $15,000

$18,000 - $3,000 = $15,000

2. $541.67

$5,000 x 12 = $60,000
$60,000 − 5,500 = $54,500

$54,500 \div 12 = \$541.67$

3. $10,325

$12,000 - \$1,675 = \$10,325$

4. $45,000

$50,000 - \$5,000 = \$45,000$
(the mortgage debt has nothing to do with this answer)

5. $10,240

$995 x 12 months = $11,940
$11,940 - \$1,700 = \$10,240$
(the property taxes are paid by the tenant and therefore are not an expense of the landlord's)

6. $146,200

$216,000 - \$45,000 - \$9,000 - \$5,000 - \$10,800 = \$146,200$

7. $120,000

4 units x 2,000 = 8,000 square feet
8,000 x $20 = $160,000 (gross rents)
Operating expenses = 8,000 square feet x $5 = $40,000
$160,000 - \$40,000 = \$120,000$

8. $52,300

$60,000 x .10 = $6,000 vacancy expense
$60,000 - \$2,000 - \$1,700 - \$4,000 = \$52,300$

9. $4,824

$600 x 12 months = $7,200 yearly gross rent
$7,200 x .08 = $576 vacancy expense
$7,200 – $576 - $1,800 = $4,824

10. $12,000
    1,500 x $10 = $15,000 gross yearly rent
$250 x 12 = $3,000 yearly expenses
$15,000 - $3,000 = $12,000

WHAT IS IT WORTH?

1. $500,000

$50,000 ÷ .10 = $500,000

2. $1,250,000

$150,000 ÷ .12 = $1,250,000

3. $923,076.92

$120,000 ÷ .13 = $923,076.92

4. $416,000

$5,200,000 x .08 = $416,000

5. 10%

$1,505,000 ÷ $150,500 = 10%
(The mortgage debt is not considered in determining the value)

6. $632,187.50 to $561,944.44

$50,575 ÷ .08 = $632,187.50
$50,575 ÷ .09 = $561,944.44

7. The 8% cap rate would give him a higher value. The lower the cap rate, the higher the value.

8. John would prefer the 9% cap rate if he were the buyer. The higher the cap rate, the lower the value.

9. Cap rates and value are like a see-saw. If the cap rate goes up, then the value goes down.

10. $655,555.55
$590,000 x .10 = $59,000 (NOI)
$59,000 ÷ .09 = $655,555.55

HOW MUCH WILL I MAKE?

1. $3,000

$11,000 - $8,000 = $3,000

2. $3,845

$15,500 - $11,655 = $3,845

3. $14,400

$800 x 12 = $9,600 (yearly mortgage debt)
$24,000 (NOI) - $9,600 = $14,400

4. $1,200 (Paul is not losing money)

$900 x 12 = $10,800 (yearly mortgage debt)
$12,000 (NOI) - $10,800 = $1,200

5. -$1,800 (Paul is losing $1,800 per year)

$1,400 x 12 = $16,800 (yearly mortgage debt)
$15,000 - $16,800 = -$1,800

6. $3,000

$1,200 - $200 = $1,000 (Property taxes and insurance have already been accounted for in the NOI, so we must subtract this from the mortgage debt)
$1,000 x 12 months = $12,000 yearly mortgage debt
$15,000 - $12,000 = $3,000

7. $14,150

$500 + $200 = $700 per month mortgage debt (we do not use taxes or insurance since they have been accounted for in the NOI)
$700 x 12 months = $8,400 yearly mortgage debt

$22,550 - $8,400 = $14,150

8. House B is the better investment

$750 x 12 months = $9,000 yearly mortgage debt
House A: $10,555 - $9,000 = $1,555
House B: $11,675 - $9,000 = $2,675

9. $42,000

$60,000 + $173,000 = $233,000 yearly mortgage debt (we do not use taxes since they have been accounted for in the NOI)
$275,000 (gross rents) - $233,000 = $42,000

10. $19,860

(we do not use taxes and insurance since they have been accounted for in the NOI)
$895 x 12 months = $10,740 yearly mortgage debt
$30,600 - $7,140 = $19,860

WHAT DO I DO NOW THAT I'VE MADE ALL THIS MONEY?

1. 27.5 years (Residential property is depreciated the same amount for 27.5 years)

2. $610.89
 $2,909 x .21 = $610.89

3. $7,051.28

$350,000 - $75,000 = $275,000
$275,000 ÷ 39 = $7,051.28

4. A house would provide a higher yearly amount of depreciation.

5. 27.5

6. $3,090.91

$120,000 - $35,000 = $85,000 (value of house, not including the land, since we don't depreciate land)
$85,000 ÷ 27.5 = $3,090.91

7. $1,421.81

$3,090.91 x .46 = $1,421.81

8. 27.5 years

9. $12,307.69

$480,000 ÷ 39 = $12.307.69

10. $4,430.77

$12,307.69 x .36 = $4,430.77

CAN I SAVE EVEN MORE MORE ON TAXES?

1. $1,281.25

$5,125 x .25 = $1,281.25

2. $12,302

$6,324 + $5,978 = $12,302

3. $8,902.25

$25,435 X .35 = $8,902.25

4. $14,000

$3,500 ÷ .25 = $14,000

5. $5,040

$14,000 X .36 = $5,040

6. $2,241.75

$8,967 x .25 = $2,241.75

7.   The principal amount that is paid on an investment loan is not deductible, because it is not an actual expense. It is merely a reduction of the amount owed on the loan.

8. $6,270.84

$17,419 x .36 = $6,270.84

9. $4,354.75

$17,419 x .25 = $4,354.75

10. $15,620

$3,905 ÷ .25 = $15,620

HOW MUCH OF THIS CASH DO I GET TO KEEP AFTER TAXES?

1. $30,000

2. $7,700

3.   Gross income – operating expenses = Net operating income

4. $173

$494.61 x .35 = $173

5. $2,967

$3,140 - $173 (from answer #4) = $2,967

A BIRD IN THE HAND

1. $86,250

$75,000 x .15 = $11,250
$11,250 + $75,000 = $86,250

2. $78,750

$75,000 x .05 = $3,750
$3,750 + $75,000 = $78,750

3.   A wise investor would choose to accept the payment up front because of the risk involved in waiting a year.

4. $275,000

$250,000 x .10 = $25,000
$25,000 + $250,000 = $275,000

5. $287,500

$250,000 x .15 = $37,500
$37,500 + $250,000 = $287,500

SCENARIO #1

1. $64,000

4,000 x $16 = $64,000

2. $59,000

$64,000 - $5,000 = $59,000

3. $655,555.55

$59,000 ÷ .09 = $655,555.55

4. $13,447.69

$655,555.55 x .80 = $524,444.44
$524,444.44 ÷ 39 years = $13,447.69 per year

5. $14,528

$59,000 (NOI from #2 above) - $44,472 (mortgage debt) = $14,528

6.   - $35,398.69 (negative amount, there is no taxable income)

$14,528 - $13,477.69 (depreciation amount from #4 above) - $36,449 (interest paid)= -$35,398.69

7.   There is no taxable income on this investment. It produces a paper loss of $35,398.69.

8. $14,528 (No taxes are owed on this cash flow)

9.   Deprecation would be higher if it were a residential investment property.

10.   -$43,398.69 (negative amount, there is no taxable income)

4000 x $14 = $56,000 gross rent
$56,000 - $5,000 (expenses) = $51,000 (NOI)
$51,000 - $44,472 (mortgage debt) = $6,528 (cash flow)
$6,528 - $13,477.69 (depreciation) - $36,449 (interest) = -$43,398.69

SCENARIO #2

1. $135,000

2,600 sf (restaurant) x $25 = $65,000
1,500 + 2,000 = 3,500 sf x $20 = $70,000
$65,000 + $70,000 = $135,000

2. $114,455

$135,000 - $20,545 = $114,455

3. $1,144,550

$114,455 ÷ .10 = $1,144,550

4.   The comparable sales are important to know. However, the value is based on the amount of income and expenses.

5. $23,477.95

$1,144,550 x .80 = $915,640
$915,640 ÷ 39 = $23,477.95